The Bird School

How to Solve Problem Behavior
with Clicker Training:

Screaming & Screeching

for Parrots and Other Birds

The Bird School

How to Solve Problem Behavior
with Clicker Training:

Screaming & Screeching

for Parrots and Other Birds

by Ann M. Castro

AdlA

The Bird School. How to Solve Problem Behavior with Clicker Training: Screaming & Screeching for Parrots and Other Birds by Ann Castro
AdlA Papageienhilfe gGmbH
Neckarstrasse 23
D-65795 Hattersheim
Germany
Find us on the web at: www.thebirdschool.com
Contact us at info@thebirdschool.com
Copyright © 2012 by Ann Castro
ISBN-13 for epub edition: 978-3-939770-68-8
ISBN-13 for print edition: 978-3-939770-66-4

Layout and photo work: Ann Castro
Front cover photo: Len Charnoff
Editor: Kimberly Collins, www.newimagedocuments.org

This book is dedicated to all the parrots and humans whose relationships no longer function. I hope I can help you with this book to find the way to each other, again.

Table of Content

1. Introduction

Screaming and its extreme form screeching are common problems in parrot keeping which particularly affect owners. A screaming parrot can be terribly distracting and wearing out the owner's patience. In addition, there are usually serious problems with unsympathetic fellow-humans.

There is pressure from family members, friends and neighbors, possibly even from the authorities, the police or animal control, to eradicate the problem. The owner is afraid of losing the animal or his home and becomes desperate.

In such a situation many parrot owners react totally inappropriately and reinforce the screaming behavior in their desperate bid to get the bird to quiet down. It is a vicious circle.

Unfortunately, a consultant is only engaged after the problem has become significant and the pressure to alleviate the issue has become immense. Thus, the owner now has a two-tiered problem which does not help the resolution:

- There is little time and leeway left for correcting the problem.
- The problem behavior is already very pronounced and established.

It would be far better, to recognize screaming behavior in its early stages and to pay meticulous attention to avoid its reinforcement. Many parrot owners, however, do not even realize in the early stages that a problem is developing.

Once they recognize that they have a problem most parrot owners at first resort to various attempts of self-help. They surf the internet, search in mailing lists, groups and forums, as well as from other parrot keepers for tips regarding the easy solution of their problem.

Unfortunately all too often these tips are totally unacceptable. They are not only counterproductive and damage the relationship between animals and keeper, but often have animal welfare relevancy. In addition, precious time is lost during which the undesired behavior instead is more strongly established. This helps neither you nor your animals! With this book I want to help you to recognize potential screaming issues early on and to ward them off.

This book should help you to remedy your birds' screaming problems yourself. To do so,

I will give you step-by-step instructions regarding the procedure that a good behavior therapist would follow to combat screaming and screeching problems.

An effective behavior therapy consists of four modules. These are usually implemented more or less in parallel:

1. Veterinary check-up
2. Optimizing the bird's keeping conditions
3. Occupation
4. Training

In this book I will explain to you what each of these modules precisely means and how you may carry them out correctly. I will be very direct and brutally honest in my communication with you. Some things I write may surprise you. At times you may even feel somewhat offended, especially as the information given may be contrary to what you have been told for many years. I do not want to upset you; however, the goal of this book is to help you effectively, by allowing you straight access to my knowledge. Only through this will you be able to help yourself, your animals, and possibly even other bird owners.

During my personal consultations, I communicate much more diplomatically, of course. But for this book to be of maximum use to you,

I must give you the best information in clear words leaving nothing open to interpretation. You need to know what I am thinking and why I resolve a certain issue in a certain way.

Before we begin, I would like to point out that the method I use for behavior therapy is clicker training. This book is an additional module to volumes I and II of my Clicker Training Series. Thus, if you are not already an experienced clicker trainer with birds, you should read both of my books, *Clicker Training for Parrots and Other Birds*, Volume I, and *More Clicker Training for Parrots and Other Birds*, Volume II.

The knowledge in those two books is required to understand the concepts in this book. Furthermore, we will use exercises from those two books for problem solving and behavior training. Thus, you need to practice those exercises with your bird first. Whenever I mention information in this book that is derived from one of the other two books, I will reference it, so you will know where you can look for more information. During behavior therapy please keep in mind that your bird's problem behavior did not appear "overnight". Similarly, it will not disappear "overnight" either. You must continuously and patiently stick with your behavior therapy in order to be successful with it. Please also take care that you do not overtax your bird with marathon training sessions. Several short training units with a pause of at least one hour in between will yield results far more quickly than one marathon session. The latter will usually lead only to frustration, lack of motivation, and annoyance. You do more damage than good with such sessions.

If you have further questions, you may post them in my discussion group or book a personal consultation with me. Information on both is given on my website (www.thebirdschool.com). I hope that soon, you too will have established happy and healthy relationships with your birds. For now, I would like to wish you much fun and success with the behavior therapy for your screaming or screeching birds.

Take care,

Ann Castro.

Ann Castro

2. Veterinary Check-up

A good behavior therapist will always insist on a thorough examination by an experienced avian vet before starting any kind of behavior consultation. This is not only beneficial for the animal and its owner, but also to protect the therapist. I urge you to diligently do likewise even if you are carrying out the behavior consultation by yourself.

The main reason is that behavior changes and problems, particularly those that appear suddenly, may be caused by health problems. The animal is in pain or feels unwell. To eliminate this as a cause or to recognize and treat it you should, without exception, see an experienced avian veterinarian when you notice any behavior changes in your birds.

I emphasize that the veterinarian must not only be specialized in birds, but that he also needs to be experienced. Proper diagnostics in parrots are difficult and an art rather than just a science. It can only be accomplished well with special training and significant amounts of experience. If you have not yet located such an avian veterinarian, you need to find one. Try asking other parrot owners. Particularly owners of very expensive kinds of parrots tend to know who the superior veterinarians are. Ask them whom they take their animals to.

In my opinion any parrot regardless of its size or monetary value deserves the best medical care you can find. It is a living being that feels and suffers. Therefore, I urge you to go exclusively to veterinarians recommended for very valuable parrots even if you only own a relatively inexpensive bird.

The second reason for my recommendation to have the bird examined by a veterinarian is that I often encounter owners who have never or extremely rarely been to the veterinarian with their birds. This may cost these animals their lives. Parrots, being prey, tend to hide their illnesses for as long as they are able to. An ill parrot will only start to show symptoms that are recognizable to most owners when he has become too weak to hide them any longer. At that point the illness has progressed to a stage in which it is much harder to cure. It may even be too late for the bird. I often hear comments from owners who claim that their parrots were totally healthy, until they suddenly and literally "fell of their perches" and died to the owners' great surprise. The problem is that these animals were not healthy. Instead, the owners simply did not notice their illness, until it was too late.

Do not judge these people. This problem is very common. These owners are not any more stupid, blind, or careless than anyone else. It is simply almost impossible to determine a parrot's illness at an early stage without the appropriate equipment and tests.

Over the years I have adopted many parrots. They always undergo thorough veterinary examinations, before they are allowed to join the flock. In all those years, I have not encountered a single bird that did not have to be treated for something or other. In several cases the birds were even carriers of lethal viral diseases that could have ended up killing my flock, if I had adopted them.

Without exception, the owners in each case were certain that their bird was healthy. They thought that I was wasting a lot of money and stressing their bird unnecessarily going through all those examinations and tests. Therefore, I implore you, please do your birds and yourself a favor and have them examined by a specialized and experienced avian veterinarian at least once per year.

The third reason for my recommendation of a veterinary check-up is that it can save your bird much suffering and even his life in an emergency. When parrots get ill they often deteriorate rapidly. It is a big advantage in such a situation to know a veterinarian whom you have already taken the bird to for an initial examination and regular well bird update check-ups. Such a veterinarian knows the animal and has its baseline information. This is invaluable in an emergency. In addition you should not underestimate how helpful it is for you, when in an emergency you already know whom you can take your bird to instead of frantically searching around for a good avian veterinarian.

3. Optimize Keeping Conditions

Behavior does not arise in a vacuum, but is impacted by several factors. Keeping conditions are one of them.

Appropriate keeping conditions significantly help in having happy and healthy animals with no behavior issues. Poor keeping conditions may turn a happy and healthy bird into a mental and physical wreck within a fairly short period of time. Please do not underestimate the importance of appropriate keeping conditions. Time and again I have seen behavior issues dissolve virtually by themselves after the keeping conditions of the animal were optimized.

Books could be written regarding the proper keeping conditions for all the different types of parrots. For our discussion, I will focus on the most relevant aspects pertaining to our topic – screeming and screeching. These problem behaviors are most severely affected by insufficient space, lack of a mate, and too little occupation.

LIFE IN THE WILD

To understand clearly how parrots ought to be kept, it behooves each of us to observe parrots living in the wild. I do understand that most of you will not be able to gallivant around some jungle for weeks at a time. Most of us will be lacking in time and funds to do so. However, there are a multitude of videos, for example on youtube.com, that show parrots in the wild.

I have set up a resource section for this book on my website (www.thebirdschool.com) where I will post links to such videos. When you view them you will quickly see that parrots are highly social and intelligent flock animals. Researchers have determined that parrots have the intelligence of three to five year old children. They also fly many miles each day and keep themselves busy with searching for food, social interaction and grooming, as well as with breeding.

Looking at those videos you will quickly realize that it is totally unnatural to keep parrots alone in a cage. Who should then be surprised that such animals start to display behavior problems sooner or later. To put it very bluntly: Keeping a parrot alone in a cage is cruel and has nothing whatsoever to do with love of animals.

WHY SINGLE KEEPING CAUSES SCREAMING

By far the most important cause for the problematic screaming of parrots is the lack of appropriate social contacts. The bird is lonely and emits contact calls. This happens because he either has no partner at all or because the human who became his partner is - not surprisingly to us - not available 24 hours per day, seven days per week. This, however, is exactly what a parrot naturally expects from his partner. The human feels irritated by these contact calls, or that he has to answer them and responds in some way or another. Regardless of whether the human calls "Hang on, I am coming", "Be quiet", or gives a treat to get the bird to shut up, any reaction to the screaming works as a reinforcer. The parrot learns, "If I scream, I receive attention". By and by the parrot expands his screaming to all kinds of situations beyond his need for social contact. He screams not only when he is lonely, but also when he is bored, frightened, hungry, or simply would like a treat.

Screaming has become an effective means for the bird to communicate with his human. Even if the human's reaction is not always the desired one or even negative, any reaction at all is still a lot better than to rot away alone and ignored in his cage.

WHY PAIRING HELPS WITH SCREAMERS

In many instances screaming issues disappear into thin air when parrots are paired with appropriate equal species and opposite gender mates. This has several reasons:

On the one hand a parrot of course loses the need for contact calls if his partner is now always close by. On the other hand a partner alleviates boredom. The pair can cuddle, preen, fight, make-up, and so forth with each other. Many birds become more active, too. They fly more for example. All this helps to exert your birds mentally and physically. This in turn counteracts any problem behaviors. Stress reduction is another important aspect of partnering parrots. A single parrot is not only bored and lonely it is also subject to permanent stress. In the wild a single parrot is a dead parrot. His partner and flock are not only important for the parrot's social interactions, but also protect him from predators. A singly kept parrot feels therefore non-stop in mortal danger. Even if the bird has become used to being alone over time, this fear will always be present subconsciously to some degree. The resulting permanent stress can contribute significantly to behavior problems. It may also cause health issues, as permanent stress weakens the immune system. Stress reaction is, therefore, an important component when pairing your parrot. None of us are able to be happy in the long run without contact to other humans, our own species. Where on earth do humans get the idea that this could possibly be different for parrots? Very clearly: It is not different at all and your animal needs an opposite gender, but same species mate.

Mini-Flock

If you have the space, I would urge you to consider keeping your parrots in a mini-flock. It is fascinating and heartwarming to observe how the dynamic changes when instead of two parrots, four or more are living together. They become much more active and adventurous. The result is animals who are mentally and physically busy and exert themselves with playing and other kinds of social interaction. This in turn reduces behavior issues significantly and also lifts the burden on the owner. You will not have to feel guilty anymore, if you are busy at work or would like to go and see a movie, instead of being at home entertaining your bird. Contrary to urban legend, the birds do not at all lose their tameness and pet quality. I have observed rather the opposite. It seems in this case also a little competition works wonders.

WHY PAIRING DOES NOT ALWAYS HELP

If the screaming problem has been in existence for a while then the behavior has been well established through reinforcement, it has become a habit.The initial reason for the screaming is no longer relevant – screaming has become an end in itself. In this situation behavior therapy must be applied in addition to the optimization of the keeping conditions to resolve the screaming problem. Through this kind of training the bird learns new behaviors that are more worthwhile to him than the screaming. At the same time the parrot learns that screaming is no longer worthwhile to him.

Detailed instructions regarding how to carry out behavior training are given to you in the relevant chapter five, Behavior Training.

How do I Pair my Parrot?

The optimization of keeping conditions is a key ingredient in the treatment of behavior issues. The most important aspect of this is the pairing of your parrot with a same species, but opposite gender mate. As this is not only of the utmost importance to your behavior therapy, but also to the well-being of your parrot, I would like to give you a brief overview as to how to go about it.

Where to Get a Partner?

I cannot stress often enough that the partner for your parrot must be same species and opposite gender in order to truly help your bird. Please do not listen to any pet stores or breeders telling you that an Amazon can be happy with an African grey or a female with another female.

It is these people's job to make money by selling animals. Thus, many of them will simply try very hard to sell you whatever "stock" they happen to have available, regardless of what is in your or your animal's best interest.

The best is to just ignore such people. A person who claims that same gender or different species pairings are appropriate is either – to be brutally honest about this – a crook or totally incompetent regarding the needs of parrots.

Both reasons make such persons unsuitable as advisers to you.

In my opinion, your best bet is to adopt a second-hand bird. This has several plus points. First of all, you are helping an animal in need which urgently needs a good home. Most second-hand birds are sexually mature adults. Thus, you know in advance what kind of personality your new feathered roommate has.

Be assured that by far not all second-hand birds are problem birds. Quite often changed life circumstances – owners dying, moving into old age homes, divorces, new partners, babies, job loss, etc. etc. – are the cause of an animal losing its home.

Secondly, when adopting a second-hand bird you avoid buying a messed-up hand-raised bird. These will often develop severe behavior issues when entering sexual maturity. Don't get me wrong. I have nothing against messed-up hand-raised birds. I love all parrots, but I don't really see the point in giving money to those people who are causing all that misery.

Thirdly, a second-hand bird will likely be a better match for your parrot from an age perspective. While it is not crucial that they both are of identical age, you should pay attention to making sure that either both birds have reached sexual maturity or neither. Parrots sold in pet stores or by breeders are often very young, thus, not suitable as a mate for your older bird.

Fourthly, previous owners are often glad to find a really good home for their feathered friend. If you are able to offer such a home, second-hand birds are often free or cost very little. The only cost arising to you would be the initial veterinary check which you would have to pay anyway. Start an emergency nest egg for your parrots with the money saved.

Fear of Problems When Pairing Parrots

Many parrot owners are fearful that their bird's pairing will not be successful or worse, that there will be serious or even lethal injuries of one or both birds. In some areas, business minded pairing stations are making money with this fear.

They suggest to inexperienced owners that the pairing of parrots is dangerous and may only be carried out by experts. Or they claim that the animal must not see its owner in order to distance itself from him emotionally and be open to his new parrot partner. Or they claim that pairing is only successful if the parrot may choose from several potential partners.

Please don't be taken in by such people. If you appropriately go about pairing your parrot, it is neither difficult nor dangerous. Not only I, but also many of my clients and forum users have successfully paired hundreds if not thousands of parrots themselves. If all these people were able to do this, then so can you.

It is my experience that parrots find most easily into each other's hearts when they feel safe. Most of my pairs that had previously been kept as single birds for many years, some even for decades, snuggled or fed each other for the first time while perching on me.

I do not recommend giving parrots away to have them paired elsewhere for several reasons. There is a huge danger for the animals to contract diseases in such pairing stations. Not every pathogen can be tested for and with such comings and goings you can be pretty certain that at some point some disease will have entered the group.

Furthermore, the whole situation is incredibly stressful for the animals. They are placed into strange surroundings with strange people and strange conspecifics. Quite often a parrot will be exposed to his own species for the first time in his life. In such a situation the bird is really more concerned about other things than "falling in love". Because of these reasons, the pairing attempts in such pairing stations can take a long time. This of course may also end up being very costly. It really is no rocket science to successfully pair parrots, as long as you pay close attention to three issues:

1. You must give the animals sufficient space and time. This allows them to approach each other at their own comfort level. Under no circumstances may you simply lock them into the same cage together.

2. Pairing should take place on neutral ground, for example your living room, and not in your original bird's territory, such as his cage, aviary, or bird room. This prevents potential territorial aggression.

3. Any indications of aggressive behavior may under no circumstances be reinforced. Owners are often overprotective and interfere already at a point where the birds are merely looking at each other sharply. The owner "consoles" or "corrects". In both cases, he is inadvertently reinforcing the unwanted behavior. At the same time he confirms to the animals that the whole situation is just "terrible". This will lead the situation to escalate and become a real problem. I hope

that experienced clicker trainers would not make such basic mistakes, but wanted to mention this for completeness sake.

The fault for failed pairings lie usually with the owners and their lack of restraint and patience. In addition, often a successful pairing is not recognized. The owner's romantic notions are not compatible with parrot reality. Sure, some parrot species are so cuddly that you can barely slide a sheet of paper in between the two birds of a pair. When they groom each other you can hardly tell where one parrot ends and the other begins. However, this is not the case for all parrot species.

African greys for example often tend to cuddle each other only for a very short time, before they go about their own business again. Most of the time the pairs do not even sleep together. Instead they play rather roughly with each other resulting in an inexperienced observer thinking that they are fighting. This isn't so.

Thus, I would recommend to let the animals carry on with their business. Give them time and space. Only interfere when there is true danger – both animals are locked together in fight. Should this really happen – and such situations are exceedingly rare, if you give the animals the space they need – you may separate the animals without any danger to them or to yourself by pouring a little bit of water over them. This usually distracts them sufficiently to let go of each other.

If you are really unsure about this, please post a question to our group and one of the experienced users or I will help.

Last but not least, I would like to remind you that the new partner must of course undergo thorough medical testing and quarantine before pairing. After all you do not want to bring any dangerous diseases into your home and to your original birds.

SPACE

Parrots must be able to mentally and physically exert themselves to remain healthy and happy. To do so they require appropriate space. Regarding the definition of species appropriate space the German government had an expert commission draw up guidelines, *The Minimum Requirements for the Keeping of Parrots*. These must be adhered to by parrot owners in order to be compliant with Germany's animal welfare laws.

In this book I will merely give you an overview table regarding the minimum cage sizes as outlined in the expert opinion. For those of you who are interested in reading the full paper, I have posted an informal translation of it to my website (www.thebirdschool.com). Please keep in mind that the minimum cage sizes are valid only in conjunction with several

hours per day time outside of the cage. According to the *Minimum Requirements,* parrots should have intensive daily flight training to keep them fit. Furthermore, the measurements describe the minimum, not the optimum size.

Most of you reading this book will not be Germans, obviously, but I find it important to show you this as an example of what appropriate cage sizes may be defined to be. They are much larger than how most people throughout the world keep their parrots.

Unfortunately for most parrots in captivity, outside the cage time often equates cuddle time with their owners. Thus they are still not moving. Nothing against cuddling. I love to cuddle with my parrots, but if in your household out of the cage time is mostly cuddle time, you must increase the time outside the cage to include flight training, or you must ensure that the space available to them while being locked-up is large enough to allow for vigorous flight. This is only possible in bird rooms or aviaries that are

Minimum Requirements
for the Keeping of Parrots

Total Body Length	Cage Measurements
Parakeets	**(Width x Depth x Height)**
< 25cm (<10")	1,0m x 0,5m x 0,5m (40" x 20" x 20")
25cm – 40cm (10" – 16")	2,0m x 1,0m x 1,0m (80" x 40" x 40")
> 40cm (> 16")	3,0m x 1,0m x 2,0m (120" x 40" x 80")
Short Tailed Parrots	
< 25cm (<10")	1,0m x 0,5m x 0,5m (40" x 20" x 20")
25cm – 40cm (10" – 16")	2,0m x 1,0m x 1,0m (80" x 40" x 40")
> 40cm (> 16")	3,0m x 1,0m x 2,0m (120" x 40" x 80")
Macaws	
< 40cm (< 16")	2,0m x 1,0m x 1,5m (80" x 40" x 60")
40cm – 60cm (16" – 24")	3,0m x 1,0m x 2,0m (120" x 40" x 80")
> 60cm (> 24")	4,0m x 2,0m x 2,0m (160" x 80" x 80")
Loris & other Fructivores	
< 20cm (< 8")	1,0m x 0,5m x 0,5m (40" x 20" x 20")
> 20cm (> 8")	2,0m x 1,0m x 1,0m (80" x 40" x 40")

significantly larger than the *Minimum Requirements* outlined.

If your birds do not engage in physical exercise in spite of being offered enough space, then you may entice them to do so with a few tricks and also through training. I will give you more information on this important topic in the next section.

Occupation

Parrots must be kept mentally and physically busy to lead happy and healthy lives. Scientists have discovered that parrots have the intelligence of three to five year old children. I am sure you can imagine how obnoxious a bored child can get.

Parrots are not different. Unfortunately, many parrot owners do not pay attention to this in the keeping of their parrots. Sadly, many parrots turn into "couch potatoes" over time, due to the limited possibilities to be active or lack of motivation. Part of our job as behavior therapists is to get those birds moving again mentally as well as physically.

Several possibilities from rearranging their habitat to actual training exist to facilitate a higher level of activity in parrots:

Arrangement of Resources

Due to misconceived welfare concerns many parrot owners place food bowls and even treats conveniently close to their parrots' favorite perches. The birds do not even have to stretch let alone really move to get to their food. This is not good. Assuming that your bird is able to fly and that there are no medical reasons that prohibit it, food and water bowls, as well as treats, toys, favorite perches and swings should be placed in the bird room or cage so that they can be reached only through flight and that the distance between them is maximized. This way your birds finally have good reasons to fly.

Brain Training

Clicker training gives you countless possibilities to train your birds' brains. Start by teaching your bird all the exercises explained in *Clicker Training for Parrots and Other Birds*, Volume I, and *More Clicker Training for Parrots and Other Birds,* Volume II, of my clicker training book series.

This ought to keep you and your birds busy for quite some time. In addition you can undertake intensive intelligence training with your birds. Parrots can learn how to count, how to recognize colors, and even how to play hide and seek.

Flight Training for Fitness

Your parrots need to exercise to stay fit. There are several methods for teaching intensive flight training exercises utilizing clicker training with your feathered companions.

Some of these I will briefly outline to you, below. If you do not know how to teach your birds these exercises, please refer to *Clicker Training for Parrots and Other Birds*, Volume I, and *More Clicker Training for Parrots and Other Birds,* Volume II, of my clicker training book series. The exercises to which I merely give an overview here are explained there in detail.

Obstacle Course

One of the first exercises you learned when you began clicker training with your birds was the obstacle course (Volume I). Usually it consists of various walking stages over various kinds of "terrain", such as branches, table tops or ropes. These paths are interrupted with various obstacles that are more or less scary to your bird. The objective is that your bird learns to climb over these objects without fear.

For the purpose of flight training you expand the obstacle course with flight stages. Instead of bridging distances with branches or ropes, your bird now has to conquer them with flight.

As always in obstacle course exercises, you lure your bird from stage to stage with the target stick.

Playing Catch

When playing catch your bird learns to fly after you. This keeps both of you on the hop. Apart from this it has the not to be underestimated effect of your parrot learning to follow you blindly. Should he ever fly away, this will make is so much easier to recover him. A recall that has been well trained through playing catch could even possibly save his life (Volume II).

I would like to add a small warning, however. If your parrot is aggressive in any way or bites, you should not train him how to play catch. After all you do not want to make it easier for him to attack you.

Retrieve

During the retrieve exercise your bird learns to bring you assorted objects. This may, of course, also be done flying (Volume II).

Search-Lost

Search-Lost is a variation of the retrieve. It kills two birds with one stone. Not only does your bird exert his brains to find the hidden object, he also applies himself physically by flying to you with the found object (Volume II).

Toys

Toys are important to keep our birds busy and challenged. Specialized stores offer a huge variety of ready-made parrot-safe toys. You can also inexpensively make wonderful toys yourself.

Parrots have different preferences when it comes to toys. Some like to destroy, some like to fiddle, yet others like noise. The toys you give your parrots should be suited to their specific requirements. At the top of the list should be foraging toys that mimic having to search for food in the wild. They also particularly help to keep your birds mentally challenged.

Toys should be exchanged frequently to keep them interesting to your birds. They should also be checked for safety at least once a day. Damaged toys can quickly become dangerous and lead to nasty accidents. Limbs can get caught in unraveling ropes or pieces of cloth, beaks can get stuck on partially opened chain links and so forth. Thus, you need to be vigilant regarding potential dangers not only when you are buying or making toys, but also during their use.

4. Behavior Analysis

Screaming may have many different triggers, such as a lack of social contacts, fear, stress, boredom, or jealousy. Often screaming is a behavior that the animal was unintentionally trained to do by unwitting reinforcement through the owner. Meanwhile, the initial reason for the screaming is not relevant anymore.

Usually, we do not know what causes the screaming as we cannot look into our birds' heads. Nevertheless, many owners feel the need to interpret their birds' behavior. They say that their bird is scolding them, because he believes, he is the boss, does not like their visitors, or similar. As a rule these interpretations are totally wrong. These owners are not only treating their animals unfairly, but – even worse – really get in the way of solving the problem.

Let's assume for example that the owner thinks his bird wants to show her that he is the boss. This belief will change the owner's attitude and behavior toward her bird. She will want to show her bird that she is the boss. As a consequence, she will start to treat her bird with less consideration. She will behave more dominantly towards her bird and maybe try to intimidate it. She may push the bird. In the worst case she may even start to abuse her bird. In short this belief is not going to help the relationship between bird and owner. Even if the bird had been a normal healthy animal before, this belief will lead to the deterioration of the relationship and to behavior issues. Now assume that this bird was screaming out of fear. He is already totally scared and screams continually, because he is afraid, and now he is besieged by his owner who wants to show him "who the boss is". The poor animal! Do you understand now how disastrous and unfair such interpretations can be?

Mind reading does not do you or your bird any favors. Therefore, please do not engage in interpretations of any kind. Your job as a behavior therapist for your bird is to objectively observe antecedent, behavior, and consequence. Then you modify the behavior by pointedly reinforcing wanted and ignoring unwanted behavior. In fact, to be absolutely precise: You are changing your own behavior in order to make it possible for your animal to change his behavior in turn.

The ABC-Analysis helps us with the observation and subsequent change of behavior.

ABC-ANALYSIS

Behavior does not develop in a vacuum. To change behavior the environment that initially caused and continues to further the behavior must be changed. Since the environment is complex, it can be difficult to identify what exactly caused the behavior and reinforced it, until it became a problem. It helps to have a system in place that allows you to collect, organize, and evaluate the required information. The ABC-analysis is such a system.

ABC stands for antecedent, behavior, and consequence. When you break down your animal's behavior into these components you will understand much more clearly what triggers and what reinforces a certain behavior. To do so you must observe the three components of the ABC-analysis in detail and capture them in writing.

Since our memory tends to adapt to our expectations, it is important to keep written notes. They help you to be objective. Written notes are unchangeable. Furthermore, they help you to identify behavior patterns. The benefit far outweighs the extra effort to write all your observations down. By the way, you should really consider keeping a journal for all your parrots' behaviors, including the desired and not just the problematic ones. All things considered, your animals consist of far more than just problem behaviors.

From history we know that the great courtesans of their time kept precise journals regarding their lovers. After all, it was of crucial importance to keep the wealthy supporter. The courtesans noted to which foods, lights, perfumes, etc. the lover reacted particularly amorously. They also noted which components had rather an adverse effect. By surrounding their lovers with only the positive, the courtesans hoped to engage their attention for as long as possible.

My suggestion to you is to treat your parrots like highly desirable lovers. Observe them in detail and take notes. You should know exactly to which foods, toys, people, colors, and other circumstances your animals react particularly positively and also to which ones they react with aversion, fear, or even aggression. This knowledge will not only help you in solving problem behaviors, but also in your daily life together. You will know how you can delight your birds and reward them. You will also know what might cause issues. This allows you to recognize small issues early on and to work on them before they turn into big problems.

The early resolution of potential problem behaviors is not only for your benefit;

you also reduce negative stress for your bird which may have a positive impact on their health and behavior. In the long run you may even save some parrots' homes for them. All too often parrots lose their homes, due to behavior issues which need not have developed in the first place and which once developed could have been resolved easily. It is such a shame and so totally unnecessary.

When working with a behavior journal I prefer a spreadsheet format. I find it gives me the best overview for later analysis. The columns for the spreadsheet are date/time, surroundings, antecedent, behavior, and consequence. Of course you may choose any format you prefer; however for those of you that would like to work with a table, there is a PDF-layout available for download or printing on my website (www. thebirdschool.com). You just need to fill in one sheet for each problem situation. You group the sheets by behavior issue and collect them in a folder to give you an organized overview of the issues. In detail, you may proceed with your ABC-analysis, as follows:

Date/Time FROM – UNTIL	Surroundings	A: What happened just before the behavior was shown?	B: Detailed description of the behavior	C: What happened right after the behavior was shown?
11/12 3.00 – 3.03 p.m.	Coco sat on his playstand in the living room. We (Mom, Grandma (visiting), Tim, Tina) were in the kitchen.	Dad laughed.	Coco screamed loudly and did not stop anymore.	Mom yelled "Shut up!" Tina ran to Coco and gave him a treat so that he would shut up (she is afraid that Coco will be given away otherwise).

A. Antecedent

Each behavior has one or more triggers. These must be identified. To do so make a note in your ABC-analysis journal of the exact situation that preceded your bird showing the undesired behavior. This includes not only what happened just before the behavior was exhibited, but also the surroundings.

Details such as time of day, weather, people present, where these people were spatially located, what these people were doing, feeding state, what was fed, etc. should all be written down in your journal.

B. Behavior

Next, you describe in detail the actual behavior that was shown. You may be surprised how complex even an apparently simple behavior turns out to be once you observe it closely and note down all the details.

Take screaming as an example. Often I hear, "My bird screams." What really happens is disclosed to me only after many questions: Is your bird's scream a short one? Does he scream only once or several times? Does he engage in continual screaming? How long does he scream for? Is there a rhythmic pattern to his screaming? Does the way he screams change during

his screaming phase? How does the screaming sound? How loud is his screaming? Does
the loudness vary? How does the bird appear
while he is screaming? Are his feathers ruffled?
What does his body posture look like? Does he
show any other behavior while he is screaming?
As a behavior therapist for your birds you must
ask all these and more questions yourself and
record your answers in detail. The better you
describe the details, the more valuable your
observations and records will be for your
behavior therapy. It is the only way for you to
accurately tell right away, whether the shown
behaviors are starting to change even slightly in
any way on account of your therapy.

C. Consequence

The situation that immediately follows the
exhibited behavior is called the consequence in
your ABC-Analysis. If the shown behavior over
time does not cease or even gets stronger you
may be certain that the consequence reinforces
it in some way.

In order for a behavior therapy to work it is
mandatory that the reinforcement of unwanted
behaviors must be stopped. But many reactions are involuntary. Often pet owners are not
aware of what they are doing. For bird owners

wanting to carry out a behavior therapy by themselves, the exact analysis of the consequence will likely be the most difficult part of the ABC-Analysis. To carry it out well, you must be highly self-critical and able to view what happens with the same emotional distance as a neutral third party observer would. If you are unsure about this point, you could ask family members or friends for assistance. Suitable persons would be those that are unbiased, great observers, and who are not reticent about being brutally honest with you.

The disadvantage of having an observer around is that his presence may change the animal's behavior. Also it can be unpleasant to let yourself be observed and criticized by family members or friends. If you are unlucky, discord – amongst the humans – may arise.

Often it works better – and has no negative effect on your personal relationships – if you film the situation with your camcorder and review it critically by yourself several times. You will be surprised about all the things that you suddenly notice that you were totally unaware of before, especially, if you watch the video repeatedly. An additional advantage of those videos is that you can pass them to experienced third parties for additional help. Remember to record all your findings from your videos also in your ABC-analysis journal.

Evaluation

After you have been carrying out your ABC-analysis for several days, you may begin with your evaluation.

Antecedents

Begin with the antecedents. Write them into a bullet list one beneath the other. For example:

- Your are leaving the room
- You are entering your home
- You are on the telephone
- You take-off your sweater

Next, consider which of these trigger you could avoid with little effort. This could be as simple as avoiding to take-off your sweater in front of your animals in the future. With the remaining triggers you will need to carry out desensitizing training. How to do so will be explained in detail in the behavior training chapter.

Consequences

Next, review your consequences. These are extremely important for successful behavior therapy, as you are reinforcing the unwanted behavior with them. Although this reinforcement happens inadvertently, it is nevertheless highly effective. Otherwise you would not be having this behavior problem, as any animal will over time only show behavior that is rewarded in some way or another.

Thus, to be successful with your behavior therapy, you absolutely must eliminate the consequences with which you are unwillingly reinforcing the unwanted behavior. I cannot stress the importance of this often enough.

Typical consequences are:

- Calling to your bird "Yes, I am coming!"
- Screaming "Be quiet!"
- Scolding
- "Calming" the bird
- Giving the bird a treat to get him to be quiet

All these examples and many others serve to reinforce the unwanted behavior that you want to eliminate. You must stop giving any of these reinforcers!

Of course there are also some consequences that have animal welfare relevancy or are just plain stupid. There are for example many parrot owners who squirt their birds with water when they are screaming. At the same time they utter their surprise to me that their bird doesn't like to shower – using the same water sprayer. I mean, what should one say to that?

Other owners bang against the cage or even throw objects at their birds. I really do hope, dear readers, that you are much more sensitive and intelligent than that and will never hurt your animals, scare them, or punish them in any way.

Our goal is to have great relationships with happy and healthy parrots. It is not our goal to abuse our animals, scare them, and to destroy even the last remnants of our relationship with them. Remember also that punishments can backfire badly and have a reinforcing effect on our birds.

5. Behavior Training

You have gone to the vets. You have started to optimize your bird's keeping conditions, and have thoroughly analyzed your bird's screaming problem. The time has come, we are finally in a position where we may begin with the actual behavior training

Our anti-screaming-training defuses particular situations to which your bird reacts with screaming through focused training. In almost all instances, a kind of desensitization takes place. Through the training, the animal learns to remain calm when confronted with certain stimuli – humans, other animals, or objects – or situations – leaving the room, entering the home – that previously caused it to scream.

BEFORE WE BEGIN

Before we begin with the actual training, I would like to give you some important pointers that you should observe during anti-screaming-training, but also when generally dealing with screaming or screeching birds. Some of these are unconscious reinforcers and some are appropriate reactions that you ought to show when your bird is screaming.

Natural Sound Emissions

Screaming does belong to all parrots normal range of behaviors. They "greet" the day at dawn and say "good-bye" to it at dusk. In the wild this serves the purpose of coordinating the flock and mates. In addition, parrots indicate through vocalizations when they have found a food source, but also to warn from danger.

Our behavior training cannot and should not serve to abolish natural behaviors. If you want to live with parrots, you simply need to get used to a certain noise level at select times.

What we do want to resolve with behavior training is to get rid of the unnatural screaming that parrots in the wild would never show. This is not natural behavior but has been trained through inadvertent reinforcement by humans.

"Be Quiet!" & Co.

Parrots love theatrics. Whenever you are yelling or scolding, this may very well be understood by your bird as a reward. With such behavior on your side, you may potentially be reinforcing the bird's screaming.

Intimidating Your Bird

On no account should you ever scare, intimidate, hit, or punish your bird in any way! You would destroy your relationship, your bird's trust in you, and his feeling safe. It simply is not worth it. If you need to let-off steam, go and beat up your pillow, if you must, go for a run or have a screaming fit in the staircase. But never, ever take it out on your bird.

He is much smaller than you. He did not choose to live in captivity. And you can be sure that he is feeling much more miserable than you are. A happy bird does not scream non-stop. Have compassion for him and his situation!

As mentioned several times already, we do not use punishment. To do so would be completely contrary to the principles of clicker training, which are based on a respectful and loving treatment with the animal.

Apart from this basic ethical attitude, punishments are not helping our cause. The animal's stress level and fear are increased through punishment and trust is reduced. This promotes screaming and other behavior problems instead of minimizing them.

Timeout

Many owners punish their birds by putting them into their cages for a timeout after they have misbehaved. You should be aware that the action of carrying the bird to his cage encompasses attention. Thus, again, you are reinforcing the unwanted behavior. Due to the time lag it is furthermore unlikely that your bird will connect timeout with his screaming. Therefore, timeouts are not effective as corrective measures. Sensitive birds may also become even more insecure through being isolated. This promotes the destructive cycle of further stress and fear and therefore screaming and screeching.

Ignore

As always in clicker training you should, even when your bird is screeching abominably, ignore unwanted behavior completely in order to not inadvertently reinforce it. This is easier said than done as screaming can be awfully wearing on one's nerves.

You may bypass this problem by investing in really good ear protection – either ear plugs or head phones. Get the industrial or professional kind. They are remarkably helpful in allowing you to ignore the screaming for as long as necessary – namely until your bird stops by himself.

Careful! Extinction Burst

When a living being has been reliably rewarded for showing a certain behavior and this reward is suddenly omitted, it will usually respond by showing an extinction burst. It will show the behavior that had previously garnered the reward in a more extreme manner. In training, extinction bursts are utilized to train shaping exercises. This is explained in detail in *Clicker Training for Parrots and Other Birds*, Volume I. However, when trying to get rid of unwanted behaviors this may, unfortunately, lead to a problem behavior intensifying before it improves.

In screaming and screeching birds, this means that your bird may scream longer and louder than before. It is of the utmost importance to stay firm when this happens. If you cave in now, your bird will have learned screaming louder and longer is worthwhile. Thus, you have a bigger problem than before. This must of course not happen! Therefore you really must ensure that no one is reinforcing the animal in any way when you are ignoring the problem behavior and your bird is showing an extinction burst because of it.

General Training

The easiest method to break screaming and other undesirable habits is general training. It strengthens the bond between you and your parrot, keeps it physically and mentally engaged, and enables him to find ways of communication apart from screaming.

I urge you therefore to teach your bird at the very least the exercises shown in *Clicker Training for Parrots and Other Birds*, Volume I in addition to your anti-screaming training. Training allows you to interact positively with your animal. The more you do so, the more your relationship will improve.

Reminder

When training, always pay attention that you do not overtax your bird. End each training session when both of you are having the most fun and give your bird a jackpot at its conclusion. This way he always has a good feeling about training. Remember, too, that your bird should be setting the pace, not you. The following exercises may take hours, days, weeks, or months, depending on your bird's personality and prior history as well as on your aptitude as a trainer.

Be patient. Be sensitive and orientate yourself not on your wishes, but solely on your bird's learning speed. This way you will accomplish the quickest and best results in training. You will hopefully live for many more years even decades with your bird. A few days more or less are hardly of importance in the big scheme of things. Give your bird the time he needs to learn his lessons.

PLEASANT SOUNDS

If your bird has any pleasant sounds whatsoever in his repertoire of sounds that he makes, you should from now on intensively reinforce those. Pleasant sounds may encompass any sounds which you find tolerable, such as singing, whistling, knocking, ringing a bell, or speaking.

In this it is completely up to you to decide which sounds you find tolerable. I, for example, do not like whistling. Therefore in my house, this sound is not reinforced in any way. Singing on the other hand, I find totally adorable, in spite of my macaws always being completely off-key. It makes me smile whenever I hear them. Thus, loud and off key singing is heavily reinforced in my house. I praise them and we have boisterous singing and dancing parties together every evening. Even new additions to the flock learn at lightning speed this way just how much fun singing is.

The utilized training method in this case is, of course, shaping (Volume I). You catch your bird showing a behavior which you deem desirable. This you reinforce through a click and treat.

Ringing the Bell

If your bird does not yet have a repertoire of desirable sounds, you may quite easily teach him to ring a bell. For this you – obviously – need a bell. Please make sure that it is neither galvanized nor made of brass. Both materials contain zinc which may lead to lethal poisoning in parrots.

If your bird is afraid of the bell, your need to do with him the approach exercise as described in *Clicker Training for Parrots and Other Birds*, Volume I, starting at comfort distance. Once your bird feels comfortable close to the bell, you may start with the actual "Ringing the Bell" exercise. Place the tip of the target stick against the bell and click and treat your bird for touching it. Repeat this several times until your bird touches the target stick which is resting against the bell without any hesitation whatsoever. Next some skill on your side is required. You must make it as easy as possible for your bird to touch the bell instead of the target stick. You could for example guide the target stick past the bell and click and treat your bird for any accidental touching of the bell.

You could also hold the target stick above the bell so that your bird has to stretch to reach it. Again, any accidental touching of the bell is reinforced with a click and treat. Experiment a bit to see how you may best accomplish this goal. Once you have repeated this several times, try to do the exercise without the aid of the target stick. It can help, if you offer the bell to your bird with your hand. If he touches it: Jackpot!

Once your bird has learned to touch the bell reliably, you may use shaping to train him to touch the bell more forcefully. The goal is for him to ring the bell loudly enough that you can hear it in the next room. Your bird should be able to call you with the bell instead of having to resort to screaming. Once your bird has realized that ringing the bell is far more reliable in calling you than screaming is, he will ring the bell instead of screaming his heart out.

Now you really should be fair to your bird and react when he rings his bell. This must not necessarily mean that you always drop everything and race to him, but you should at least acknowledge it in some way. You could call "I am coming" or "wait, not now". Most parrots understand the meanings of these different answers quite rapidly and accommodate themselves accordingly.

Screaming When Leaving the Room

A common trigger for screaming is when the owner leaves the room in which his bird is located. This can be easily remedied with a bit of training. You must begin by determining the exact point at which your bird begins to scream. Ideally you have already determined this during your ABC-Analysis. If not, you must find this out before beginning with this exercise. Does your bird start to scream when you turn away from him? When you open the door? Or does he only start to scream once you have moved out of his sight? To find out position yourself directly in front of your bird. Then start moving away from him exceedingly slowly. Make a note of when exactly he starts to scream.

Wait, until he stops screaming again and repeat this exercise. After several repetitions you should know fairly precisely when your bird starts screaming. Let's assume that this is when you are opening the room door with your back turned towards your bird. This distance less two steps is what we will call the distance point for this anti-screaming-exercise.

The Walk to the Room Door

The starting point for this exercise is right in front of your bird. Start walking away from there, as if you wanted to leave the room, but stop at the previously determined distance point. If you have correctly determined this point, your bird should still be silent. Click him for this and give him a treat. As you have to return to the bird to give him the treat you should be back at the starting point now.

Walk again to the distance point. Your bird should still remian relaxed and quiet inside his cage. Reward him again with a click and treat.

Again, walk away from your bird. This time, however, proceed one step which is approximately 12 inches, beyond the distance point. If your bird still remains quiet at this new distance point, click and treat him for that. Repeat the exercise with the new distance point

When you have repeated the exercise for this distance enough times that your bird is not only quiet, but also utterly relaxed, go another 12-inch-step further. In our example this would take you right in front of the room door.

If you have not reached the door, yet, keep repeating the previous steps, until you get there. Should your bird start to scream at any point during the exercise, you must wait, until he is completely quiet again. Then and only then may you click and treat him. Repeat the exercises at the last distance at which your bird still remained totally quiet as often as necessary, until he remains totally relaxed while you move back and forth between him and the distance point. Once you have achieved that, you may go a few inches beyond the distance point during the next repetition of the exercise. Do not go a whole step, it would probably overtax your bird at this point. Many smaller increases will get you results far quicker than one big step that is more than your bird is able to handle. If your bird remains silent, click and treat him. Repeat the exercise at this new distance point again as often as required for him to remain totally comfortable. Then inch another little bit forward.

Keep going like this until you can walk from your bird to the door without any problems. Should your bird start to scream during any part of the exercise, follow the procedure outlined above by reverting to the last training step at which your bird was still totally relaxed and quiet. Please pay particular attention during this training that you never reinforce your bird in any way should be begin to scream.

Through the Door

The next training step is to practice your going through the door. Start by breaking this action down into as many little ministeps as possible. These you practice one after the other, until you are able to carry out the entire action sequence without your bird starting to scream. The ministeps could be, for example:

- Lifting your hand
- Placing your hand on the doorknob
- Turning the doorknob
- Opening the door
- Moving towards the threshold
- Stepping over the threshold
- Closing the door behind you

The actions of opening and closing the door may also be subdivided in many little one inch steps. Now you may practice every single one of the steps, until your bird remains totally calm for every single one of them. He should not just not scream, but also be totally relaxed before you proceed to whatever step comes next.

When you have completed the final step – closing the door behind you – don't wait, but immediately click and return to your bird.

Chaining the Total Sequence

Once you have trained each step of the leaving-the-room sequence thoroughly enough for your bird to remain completely relaxed and quiet, you may start training the individual steps in series.

Start by standing in front of your bird and then walking slowly, but without pauses, away from him and out of the room. Then close the door. If your bird remains totally relaxed and calm, click and treat him for this.

Should your bird by any chance start screaming at any point then you will need to practice that particular step more thoroughly, before you try to chain the steps, again.

If there is no problem whatsoever you may increase the level of difficulty by increasing the speed at which you are able to leave the room at each repetition. In the end you should be able to virtually race away from your bird while he remains completely calm and quiet.

Practicing Absence

Once you are able to leave the room without any problems whatsoever and your bird does not begin to scream, you may begin to practice your absence.

During this exercise you had so far opened the door right away again after closing it, walked to your bird and gave him his treat. Now it is time to practice the door remaining closed for longer and longer time spans. Begin small, closing the door only for one second. Then proceed to two seconds, three seconds and so forth.

In the end you should be able to leave the door closed for several minutes without your bird starting to scream.

Increasing the Degree of Difficulty

You may raise the level of difficulty further by starting to speak outside of your bird's room, turn on a radio, have a conversation or make noise in any other way. Again, you must proceed in ministeps. You could start, for example, by saying just one word, then two, then three, and so forth. Remember to keep clicking and treating your bird for remaining quiet.

ENTERING THE HOME

Many parrot owners have established the ritual of letting the bird out of his cage, or feeding him, or both right after arriving home. Therefore many birds impatiently await the return of their humans and "greet" him impatiently and loudly, as soon as he approaches the home.

In general the owner races to the animals as quickly as she can to cut-off the screaming. Thus she rewards her birds for all that screaming. Apart from the fact that it is awfully stressful for the owner to be confronted by a screaming mob the second she walks through that door, it can also lead to serious problems with the neighbors. Therefore, the "Entering the Home" exercise should be thoroughly trained, as well. Basically, this exercise is the complete reversal of the "Leaving the Room" exercise. What makes this exercise much harder, though is that it is more difficult to click and treat the bird appropriately when entering the home. You really will need some patience and strong nerves.

The goal of this exercise is of course to be able to approach and enter your home and go to your bird without him starting to scream. During practice, this means that every time your bird starts screaming you must wait, until he is quiet again, before you may proceed further in his direction.

This could mean that you may have to spend a lot of time waiting, before you are actually able to enter your home and go to him. Prepare yourself well to make this waiting more tolerable. Grab a good book, some chocolate, a bottle of red wine, or whatever will help you to whittle away the time more pleasantly. Make sure you go to the bathroom before you begin. Ear plugs for you and your neighbors would be advisable, too. If you have a halfway decent relationship with your neighbors I would urge you to inform them beforehand of your plans. They most likely have already noticed that your bird screams whenever you come home. Thus, they should be pleased that you are starting to do something about this problem, now.

As your neighbors are likely not experienced animal trainers or psychologists, you really need to explain to them that there is the possibility of the screaming getting initially worse, before it stops altogether. Who knows, if you are lucky, maybe they'll keep you company during the wait.

Start with the walk from the street to your front door. As soon, as you hear your parrots scream, stop completely and do nothing, absolutely nothing – except drink red wine, eat chocolate, etc. –, until your birds are quiet once again. Only then may you proceed in your approach. Whenever your animals start to scream, stop. Continue only after your birds are quiet, again. Proceed in this way slowly and with much patience, until you are finally standing in front of your birds' cage and can let them out. You will need to proceed like this now day-by-day in all situations in which your birds scream, because they are impatient. Screaming may very simply never ever be rewarded. In contrast you may be generous with surprise treats and attention whenever your birds are quiet. Over time your animals will learn that screaming is simply not worthwhile. Silence, however, is.

THE TELEPHONE

Many parrot owners find it almost impossible to have a telephone conversation from home, as their feathered companions loudly "join" in. Of course you can and should also train your birds to be quiet while you are on the phone. To do so, you fake a telephone conversation. Hold the telephone to your ear, as if you were having a phone call, but do not say anything, yet. If your birds remain silent and relaxed, click and treat them. If they scream, ignore them.

If your birds remained silent, you may proceed by speaking one word , for example "hello", into the phone. If your birds remain quiet, click and treat them. Should they start to scream, ignore them completely without saying a word. Once they are quiet again repeat the exercise. Keep repeating the "hello" step several times, until your birds remain totally relaxed and quiet. Then you may slowly start to increase the length of your fictitious phone call. This means that in the next repetition, you may say two words instead of one. This too will be rewarded with a click and treat, if your birds remain silent. Repeat this step also several times until your birds are completely relaxed and quiet. Then say three words. Again your birds are clicked and treated for remaining silent.

In this way you may slowly progress until you can have a longer, calm fake conversation. If you are able to simulate such a conversation on the phone with your birds remaining totally quiet, you may increase the level of difficulty by allowing your voice to become increasingly more lively during the fake conversation and later on also louder.

After you have practiced both thoroughly, it is time to test your progress with a real telephone call. This situation is often somewhat different for you and your birds, because your attention is clearly not focused on them anymore. The procedure is the same as for the fake conversions we used before. Please choose a person for this exercise as a conversation partner that you have previously briefed and who understands that you may need to interrupt the conversation frequently. Practice a real conversation as often as necessary for your birds to remain quiet no matter how animated you talk, how loud you laugh, or how long the phone call is.

Sounds

Oftentimes sounds are the triggers for a parrot's screaming behavior. Here too, desensitization can work miracles. During desensitization training you accustom your bird gradually to the sound by letting him hear it for increasingly longer periods of time and afterwards in increasing volumes.

Let's assume that your bird reacts to dog barking with intense screaming. To get your bird used to dog barking you will need a recording of this sound. The training will be particularly authentic, if you can record the dog barking that normally sets your bird off. This could be, for example, your own dog's barking or the neighbor's.

If this is not possible for you, it is still possible to work with "canned" sounds. Some of these are available free of charge on the internet or they may be purchased as sound-samplers. To enable you to start with your training right away, I have uploaded various sounds to the resource section for this book on my website. These uploads include sounds, such as dog barking, sirens and telephone ringing. I have also posted a link there to a free-of-charge sound website on which you will be able to find many more sounds to practice with.

Begin the exercise with the volume of your media player at a very low setting. Play the sound that normally elicits screaming to your bird for approximately a second. If your parrot remains calm and relaxed, click and treat him. Next, play the sound at the same low setting for two seconds. If your bird still continues to be quiet and relaxed, click and treat him again. As always repeat the exercise at each training step as often as necessary for your bird to remain totally calm and relaxed. Only then should you proceed to the next level.

Continue like that, until you are able to play the sound for approximately five seconds. Keep repeating this exercise several times at that level, until you are truly satisfied that your bird does indeed remain completely calm and relaxed. When you have achieved this, you may progress to the next part of the exercise in which you begin to gradually increase the volume. This increase should be absolutely minimal to ensure that your bird is neither startled nor begins to scream. If that happens, it means that you have increased the volume too much. In that case you must go back one or more steps in your training, until you are again at a level at which your bird remains totally calm. Then you slowly inch forward again – more cautiously than before.

In most cases, though, as long as the volume increase is absolutely minimal, it should present no problem to play the sound for the full five seconds. If this is the case, click and treat your bird and then proceed to notch up the volume another little bit.

Remember to continually observe your bird's body language. Should he become tense, you need to remain at that level and repeat the exercise as often as necessary for your bird to be able to handle it while remaining completely relaxed and without beginning to scream.

Continue to proceed in this way, until you can play the sound at the full volume at which it normally occurs for five seconds. When your bird handles this well, you may gradually increase the play time for the sound, as explained in the first part of this exercise, until you are able to play the sound at full volume for as long as your bird would normally be exposed to it.

SCREAMING TRIGGERING SITUATIONS

A good portion of anti-screaming-training consists of training situations to which your bird would normally react with screaming. The final training goal of all these exercises is for your bird to remain calm and relaxed even when confronted with situations to which he previously reacted with screaming. Which situations you need to practice with your birds will be different for each one of you and has been determined through the results of your ABC-Analysis. The procedure, however, is always the same. Start by splitting the screaming triggering situation into as many steps as possible. The smaller those steps are, the easier and faster you will gain successes with your training efforts. Under no circumstances may you overtax your bird by making individual training steps too large. This would only lead to failure!

After identifying all of the training steps, you need to practice each one of them separately, until your bird remains fully relaxed for that particular step and can be clicked and rewarded by you for this.

When your bird has learned to stay relaxed for each one of those mini-steps, you may begin to gradually increase the speed with which you go through the steps and subsequently start grouping them together into increasingly larger training steps. The final goal is to have chained all those steps together into one coherent behavior sequence which you may go through quickly and without any screaming reaction whatsoever by your bird.

Example: Taking-off the Sweater

Many diverse situations may be triggers for screaming behavior. It is likely that the situation startles or frightens the animal. Again we will practice step-by-step in order to desensitize the bird with regards to the situation.

Let's use as an example the taking-off of a sweater. Many parrots react strongly to this: hands are gesticulating wildly, the face of the known and trusted person cannot be seen anymore and this scary thing moves about as if it had a life of its own.

Deconstruct Into Tiniest Possible Steps

Think in detail about how you, a roommate or a visitor take of their sweaters. There are many different ways of doing this. Some people take hold of the lower seam with their hands crosswise and pull the sweater off over their head in one smooth move. Other people first pull their arms out of the sleeves, before they grab the neck opening from the inside and carefully lift it over their head. Then there are people who pull out one arm first, then the head, and then the other arm.

Isn't it remarkable, how such a simple action may be done in so many different ways? If your bird screams only when certain people take-off their sweaters but not for others, then this could be the reason. Observe also, at which stage of the taking-off the sweater your bird starts to scream. Is it right at the beginning of the action or only at the end, when the sweater is thrown across the room into the laundry basket?

As you see, a good behavior therapist needs to observe accurately. If those kinds of details are lacking in your ABC-Analysis, you need to go back and redo it and also its evaluation. Only then will you be able to clearly identify which little step precisely it is that your bird is reacting to with screaming and needs to be trained to remain relaxed for. Let's assume that taking-off your sweater comprises of the following actions for you and that your bird reacts to each one of these steps equally with screaming:

- You lift your left hand.
- You move your left hand in front of your body.
- You lift your right hand.
- You move your right hand in front of your body.
- You grab the right sleeve cuff with your left hand. Both hands are in front of your body.
- You pull the cuff with your left hand. At the same time you move your right hand and arm in the opposite direction.

- You pull your right arm out of the sleeve.
- You lift the neckline over your head with both hands.
- You lift the right hand.
- You move the right hand in front of your body.
- You lift the left hand.
- You move the left hand in front of your body.
- You grab the left sleeve's cuff with your right hand.
- You pull the left sleeve off the left arm.
- You hold the sweater in your right hand.
- You grab the sweater also with your left hand.
- You shake the sweater out.
- You fold the sweater in the air.

This is approximately what deconstructing your taking-off the sweater into tiny steps could look like. In training, all actions that comprise a directional movement will be further broken down into a multitude of one-inch steps.

Each action will be trained individually in many mini-steps during our anti-screaming training. During each one of those mini-steps you observe your bird closely and pay particular attention that he is calm and relaxed before you reinforce him by clicking and giving him a treat.

As in the previous exercises, this is followed by going through the mini-steps successively faster before you finally start to combine them into increasingly larger steps. In the end you should be able to run through the entire action sequence quickly and smoothly while your bird remains relaxed, and most importantly, does not start to scream.

Once you have accomplished this, you can start with the transfer exercises. Train the action of taking-off the sweater in different locations and also with different people.

In this way you are able to defuse any screaming triggering situation. Just follow step-by-step the methodology that I have outlined to you here. Always take care to proceed in mini-steps without overtaxing your bird.

TRICKS FOR NATURAL BEHAVIORS

The main goal of any behavior therapy is to correct learned unwanted behaviors. It is not and never should not be, the goal of behavior training to eliminate natural or normal behaviors. Nevertheless certain normal behaviors can still be a problem within a certain surrounding, for example, if your parrots are of the opinion that the day should be greeted at five o'clock in the morning. Even if you as a parrot owner are fully understanding of this, there is a good chance that not all of your neighbors will be likewise understanding.

To stop parrots from saying "hello" and "good-bye" to the day is not an option. What is to be done, then? Luckily there are several tricks which aid parrot owners in this situation:

Darkening the Room

Outfit your bird room with lightproof drapes or shutters. This allows you to decide when the day begins and ends enabling you to move the greeting orchestra to a time of day that is more humane to your neighbors than five a.m.

I would not recommend to cover the cage, because a cage that may be easily covered is way to small for most kinds of parrots, does not comply with species appropriate keeping and therefore has animal welfare relevancy. Thus, I am assuming, dear reader, that this is not an option for you, anyway!

Shortening Dusk and Dawn

You may shorten the dusk and dawn phases through the use of artificial lighting. The short-ened twilight phases give less opportunity for extended noise orgies.

You may question whether this is compliant with species appropriate keeping. Normally, I recommend to keep parrots as naturally as possible. However, what is natural for tropical animals? Surely not fifteen hour long summer days or extended twilight phases. In the tropics days and nights are approximately equally long and dusk and dawn phases are very short. Thus one could well argue that using darkening and lighting equipment allows for the adaption to the natural settings in our parrots' home regions.

Apart from that I believe it is legitimate to attempt to avoid stress with neighbors or even the loss of your home. A shortened twilight phase must be surely the lesser of the two evils, especially as these kinds of situations often result in the abandonment of the animals.

Occupation

You can also try to keep your parrots busy during the twilight phase. This will not work for all parrots, but it is worth a try. You could for example have your clicker training sessions during the twilight phase, but also freeflight (inside your home) or cuddle times.

Another option is to feed your birds during twilight phases or give them special treats. They are then so occupied with eating that they forget

79

to scream. Once they are done eating the twilight phase has already passed.

Pair Keeping

The importance of pair keeping to reduce screaming has already been discussed in Chapter 3, Optimize Keeping Conditions. Please read, if you have not already done so.

6. Closing Remarks

The success of anti-screaming training depends completely on the consistency of the trainer to catch and reward quiet behavior while completely ignoring any screaming. This should become a part of your regular life and not only observed during training sessions.

The methodology of anti-screaming-training is always the same: The trainer progresses in mini-steps giving him the chance to catch his birds before they start screaming in situations in which they would normally scream. The animals are then rewarded for remaining quiet and relaxed during each of those mini-steps.

The examples given in this book may be applied to any screaming situation. Just remember to always deconstruct the situation into smallest possible steps and to practice those individually. Even if screaming parrots can be extremely annoying, please, always keep in mind that the bird is not your enemy. He is at least as confused, hurt, lost, and unhappy as you are. He is a victim of his circumstances. He certainly did not choose to live in captivity. We are forcing him. Ethical considerations aside, for this reason alone we should always meet our animals with friendliness, love, and understanding.

You also should always keep in mind that you are the adult in this relationship. It is up to you to resolve any problems in order for you and your feathered roommates to find the road back to each other and to happy and healthy relationships.

If you have any questions after reading and working through my books, I am available for private consultations. Alternatively, time permitting, I also offer advice in my online parrot group. Information on both is given on my website (www.thebirdschool.com) where you may also read articles on various parrot related topics, as well as a blog.

I would be delighted to hear how you and your feathered friends fare in your training efforts. If you would like to mail me your story, photos, or videos I will gladly post them on my website as an inspiration to other parrot owners.

But for now, I would like to simply wish you much success with your training.

Take care,

Ann Castro.

Who is AdlA Papageienhilfe gGmbH?

"Papageienhilfe" is the German word for parrot aid. AdlA stands for "Amigos de las Aves" which means friends of the birds. AdlA Papageienhilfe gGmbH is a not for profit organization whose objective is to help parrots and their owners.

Our goal is to reduce the number of parrots that need to be rehomed by giving thorough advice before and after the acquisition of parrots. This encompasses giving information to parrot owners regarding species characteristics, appropriate keeping conditions, and how to create or optimize them in their own homes. In addition, we offer behavioral therapy for birds with problem behaviors.

Our consultations comprise:

- Purchase advice regarding different species, methods of raising and procurement, challenges and obligations to the owner. Where possible and appropriate we try to place adoption birds.

- Optimization of keeping conditions, such as cage sizes, mates, nutrition, light, and humidity
- Advice regarding required well bird health checks, particularly initial and yearly check-ups, as well as quarantine procedures
- Emergency counseling: First aid to sta-bilize the bird until he can be seen by an avian veterinarian
- Behavioral therapy and training

Even if it is our primary goal to assist the own-ers to prevent birds needing to be rehomed, it is not always possible. In those cases where a bird needs to find a good new home, we assist in placing the bird out for adoption. Sometimes, when a bird is too handicapped or mentally troubled to be placed, we integrate it into our flock.

Our work is financed through donations:

TaunusSparkasse Hoechst

Banking Account No:

DE89512500000000 320382

IBANBIC (former SWIFT):

HELADEF1TSK

Paypal: papageienberatung@googlemail.com

The Author

Since childhood, Ann Castro has been sur-
rounded by birds and other animals. Both parents
being medical doctors – her father a psychiatrist
and neurologist, her mother a general practitio-
ner who also bred budgerigars – Ann was intro-
duced to behavioral, as well as medical topics
at an early age. The author has been involved
with parrots for many years. She teaches clicker
training for birds and gives advice to parrots
owners regarding all issues related to their pets.
Her area of specialization is the resocialization
of birds with behavioral issues.

In 2003, Ann Castro founded a not-for-profit
organization, the AdlA Papageienhilfe gGmbH.
AdlA stands for Amigos de las Aves which
means friends of the birds. Papageienhilfe is
the German word for parrot aid. The aim of
the AdlA Papageienhilfe is to improve the relationships and understanding between birds and
owners and thus minimizing the number of
birds losing their homes. Ann Castro's advice

is sought by pet owners, veterinaries and pet shops alike. She has also appeared in various TV shows as a parrot expert.

Ann Castro, a born Canadian who currently lives in Germany, holds a Bachelor of Applied Science degree in Chemical Engineering from the University of Toronto and a Masters degree in Business Administration from York University, Canada.

Other Publications by Ann Castro

The Bird School (Volume I)
Clicker Training for Parrots and Other Birds
This book for beginners shows you how to do clicker training with your birds. You will learn how to teach your parrots basic obedience, but also just for fun exercises. Clicker training is entertaining for birds and owners and enhances the bird-human relationship.

Topics covered are: Up, Down, Come, Let Go, Leave It, Go Away, Relaxation Exercises, Obstacle Course, Turning in a circle, Loop de Loop, Nodding, Wiggling the Tail, How to Tame a Bird
152 pages, ISBN-13: 978-3939770039

The Bird School (Volume II)
More Clicker Training for Parrots and Other Birds
Volume II continues where Volume I left off. Learn how to teach your parrot advanced obedience exercises, fun games you can play together, as well as

various exercises for grooming, emergencies, and medical care.

Topics covered include: Learning the name, to your spot, stay, into the cage, into the carrier, drinking from the syringe, shake hands, filing the toenails, wave, retrieve, lost-find, playing catch, big eagle, the head grip and the well game.

224 pages, ISBN-13: 978-3939770596

The Bird School (Volume III)
Biting & Aggressions: How to Solve Problem Behavior with Clicker Training

A behavior therapy book for experienced clicker trainers. This book will teach you how to resolve biting and aggression problems with your birds using clicker training.

100 pages, ISBN-13: 978-3939770619

As Anna G. Shiney

Clicker Training the Law of Attraction.
How to Treat the Universe Like a Dog

Applying the principles of clicker training to make the law of attraction work for you.

www.annashiney.com

ISBN-13: 978-3939770275